Spiders

The illustrated identifier to over 90 species

Spiders

The illustrated identifier to over 90 species

Ken Preston-Mafham

APPLE

A QUINTET BOOK

Published by The Apple Press
6 Blundell Street
London N7 9BH

ISBN 1-85076-832-3

This book was designed and produced by
Quintet Publishing Limited
6 Blundell Street
London N7 9BH

Creative Director: Richard Dewing
Art Director: Clare Reynolds
Designer: Roger Fawcett-Tang
Project Editor: Doreen Palamartschuk
Editor: Maggie O'Hanlon
Illustrator: Tony Oliver

Typeset in Great Britain by
Central Southern Typesetters, Eastbourne
Manufactured in Singapore by Eray Scan Pte Ltd
Printed in Singapore by Star Standard Industries Pte Ltd

All photography, including jacket, reproduced by permission
of Ron Brown, Ken Preston-Mafham, Mark Preston-Mafham
and Dr. Rod Preston-Mafham for Premaphotos Wildlife.

Contents

Introduction   **6**

What is a spider?   *6*

The spider's body   *8*

How to use this book   *10*

Classification of spiders   *10*

Spider identifier   **12**

Index   **80**

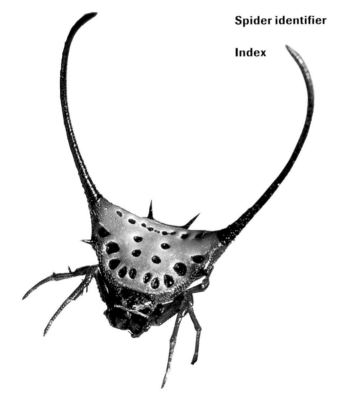

Introduction

Spiders are everywhere. Every garden or backyard will be full of them, no matter how carefully manicured it may be. Houses are the preferred abode for several kinds of highly urban spiders, while the spider inhabitants of a grassy meadow will be numbered in their millions. Ponds and lakes have their quota and, at the other extreme, deserts are no bar to spiders, which often escape the drought by living in cool burrows deep beneath the sun-baked surface. Even the inhospitable slopes of Mount Everest at an altitude of over 20,000ft/6,000m are home to spiders that live nowhere else on Earth.

The number of spider species so far described is just under 40,000, and this is estimated to represent perhaps one quarter of the real total. In Europe, where the tradition of collecting and studying spiders goes back a long way, the total just exceeds 3,000 species. In North America the total is currently rather smaller, but new species are constantly being described.

What is a Spider?
It is quite common to hear spiders being referred to as insects. This is quite wrong. Apart from the fact that insects and spiders are both arthropods, they have little in

Harvestman.

Velvet mite.

Scorpion.

common. Arthropods are animals which have jointed legs and a hard outer skeleton, called an exoskeleton. As the spider or other arthropod grows, it sheds this hard exoskeleton in a series of molts. Insects always have a pair of antennae mounted on the front of the head, their body is divided into three parts (head, thorax, and abdomen), and they have six legs. Spiders never have antennae, they have eight legs, and their body is divided into only two parts: a fused head and thorax called the cephalothorax (or prosoma), and the abdomen (the opisthosoma).

Spiders are members of the class **Arachnida**, which includes some other rather spider-like creatures. The **harvestmen**, belonging to the order Opiliones, generally have long, spindly legs and rather small, rounded bodies. The attachment of the cephalothorax to the abdomen is broad, whereas in spiders there is a narrow stalk. Harvestmen only have two eyes, often perched on turrets, but most

spiders have eight eyes, although some only have six, and a few rare ones only two (but not on "turrets"). **Mites** and **ticks** (order Acari) are mostly very tiny, and it is almost impossible to distinguish the division of the body into head, thorax, and abdomen. **Scorpions** (order Scorpiones) can easily be recognized by their long tails with the stinger at the tip and their large, pincered front legs. **Windscorpions** (order Solifugae) are the sprinters of the arachnid world, running on only six of their eight legs and using the front pair as feelers. **Whipscorpions** (order Uropygi) number fewer than 100 species

Windscorpion.

Whipscorpion, or vinegaroon.

and also walk on just six of their eight legs. The abdomen is clearly divided into segments. They usually have pincer-like pedipalps and long, whiplike tails, although they cannot sting. **Tailless whipscorpions** (order Amblypygi) lack the tail and are more flattened than whipscorpions. The head and thorax are broad, and the abdomen is attached by a stalk. All the legs are slender and held out to the sides, but the front pair is very thin and whiplike, and is not used for walking.

The Spider's Body

As already mentioned, the body is divided into two sections: the **cephalothorax** and the **abdomen**. The top of the cephalothorax is covered by a hard plate called the **carapace**. The front of the carapace bears the group of eyes. These are simple in structure and shine brightly in the light from a flashlight, which is a good way of finding spiders at night. Below the head lie the **jaws**, or **chelicerae**. These are formed of two sections: a stout, fixed basal section tipped by a movable **fang**. In most spiders this can dispense venom through a duct leading from a venom gland. The **mouth** lies below, beneath a lower lip, or **labium**. Spiders dribble copious quantities of digestive juices onto their food (all spiders are carnivores), turning it into a kind of broth. In addition, the basal part of the chelicerae in many spiders is equipped with

Tailless whipscorpion feeding on a cockroach.

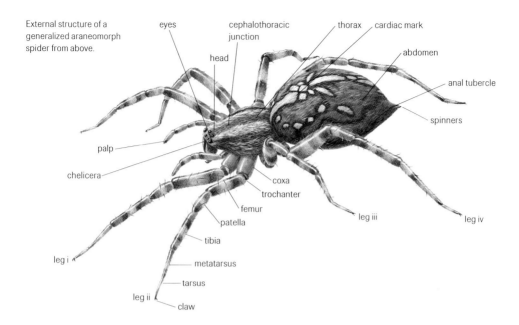

External structure of a generalized araneomorph spider from above.

eyes
cephalothoracic junction
thorax
cardiac mark
head
abdomen
anal tubercle
spinners
palp
chelicera
coxa
trochanter
femur
leg iii
leg iv
patella
tibia
leg i
metatarsus
tarsus
leg ii
claw

teeth that can grind up the prey into an unrecognizable pulp. This is then sucked back up by the spider, using the powerful pumping action of the stomach. Some spiders (e.g. crab spiders, family Thomisidae) pump the digestive juices into their prey and extract the resulting broth from inside, leaving a virtually intact exoskeleton.

Between the jaws and the first pair of legs lie the **pedipalps** (often referred to as the **palps**). These are jointed, and in females and juvenile spiders they are slim and rather resemble small legs. In male spiders the tip of the pedipalps is greatly enlarged and often complex. Before going in search of a female, the male builds a small web on which he deposits a droplet of sperm from the orifice at the rear of the abdomen (**anal tubercle**). He then sucks the sperm back up into the pedipalps. During mating these are

then ready for inserting into the female's genital pore (**epigynum**).

The abdomen is relatively soft and flexible. Near its tip lie the **spinners**, which produce various types of silk, depending on its final use. This could be for the external coating of an egg-sac, the wrapping of prey or the spiral strands of a web. Most spiders constantly trail a thin dragline from one of the spinners, and this functions as a kind of safety rope. Silk starts out as a liquid from special glands, but it quickly solidifies as it is stretched. In some spiders (called the cribellates) a flattened, sieve-like plate called a **cribellum** lies just in front of the spinners. By pulling a special row of spines (the **calimastrum**) on the rear leg through the cribellum, very fluffy, multistranded silk, often called **hackled-band silk**, is produced via the spinners.

How To Use This Book

The spiders in the identifier section of this book are arranged according to their classification in two groups: the suborders Mygalomorphae and Araneomorphae, which both fall within the order Araneae. A brief description introduces each family, and for each species representing the family the species name and common name are given, together with information on its general characteristics, and known distribution.

When you see a spider, note its surroundings: such as whether or not it is in a web, on the ground, or on a leaf. Consult the information under the family heading and then try to match your specimen to one of the photographs, and the descriptions of the spider's appearance, and habits. This will help you decide whether or not you have chosen the correct genus or species. Lengths given are of the body only and do not include the legs.

Classification of Spiders

Spiders are classified in accordance with certain rules governing the whole of the animal kingdom. This involves an hierarchy, starting with the major grouping, and ending up with the smallest indivisible unit, the species. If we take as an example the garden spider (*Araneus diadematus*), the classification would be as follows.

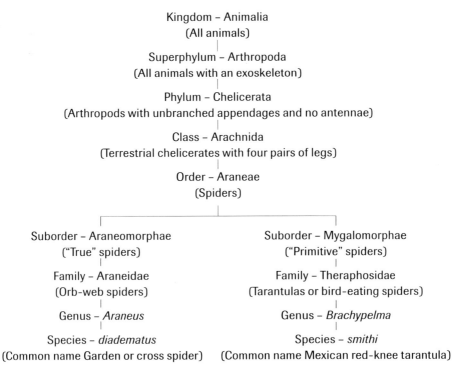

Kingdom – Animalia
(All animals)

Superphylum – Arthropoda
(All animals with an exoskeleton)

Phylum – Chelicerata
(Arthropods with unbranched appendages and no antennae)

Class – Arachnida
(Terrestrial chelicerates with four pairs of legs)

Order – Araneae
(Spiders)

Suborder – Araneomorphae
("True" spiders)

Family – Araneidae
(Orb-web spiders)

Genus – *Araneus*

Species – *diadematus*
(Common name Garden or cross spider)

Suborder – Mygalomorphae
("Primitive" spiders)

Family – Theraphosidae
(Tarantulas or bird-eating spiders)

Genus – *Brachypelma*

Species – *smithi*
(Common name Mexican red-knee tarantula)

The symbols given below accompany each entry and are intended to give vital information about the habits of each spider, and the differences between male and female.

Where is the spider found?

 In open ground such as grasslands, deserts, heaths, and moors.

 In forest and woodland, but often in open spots and along roads.

 Common in gardens.

 Mostly in buildings.

 Mostly in or beside water.

Where is the spider most likely to be sitting?

 In a sheet-web.

 In a silken tube or cell, often under a stone or log, or partly in the ground.

 Inside a burrow.

 On the ground.

 On leaves.

 In an orb-web.

 On the bark of a tree.

 On a wall or rock.

 On flowers.

 In a tangle- or scaffold-web.

Are the males and females very similar in size and color?

 Very similar.

 A little bit different, mainly in pattern and color, with males being smaller.

 Very different. Males are nothing like females, and could be mistaken for a different species.

Suborder Mygalomorphae

*The spiders in this suborder are less specialized than those in the
suborder Araneomorphae, which are regarded as more advanced.
The most significant single difference between the two suborders is the
arrangement of the jaws. In the mygalomorphs these have an up-and-
down action, moving parallel with the front-to-back axis of the body.*

Purse-Web Spiders *Family Atypidae*
This is a small family of stoutly built, shiny-bodied spiders that live in silken tubes. There is a
line of sharp teeth along the rim of the basal cheliceral segment, against which the fangs close.
This enables the spider to snip holes in the tough silk of the sealed tube and draw prey inside.

ATYPUS AFFINIS

COMMON NAME Purse-web spider
DESCRIPTION The $^{11}/_{16}$ in-/17 mm-long female
(illustrated) is a glossy, fat brown spider
with very short, stubby legs and huge,
projecting jaws. She usually spends her
entire life within a closely woven, silken
tube resembling the finger of a glove. This
tube lies on the ground among grass or
under a large stone. When an insect strays
onto the silk, it is speared on the spider's
long fangs, which stab "blind," but with
great accuracy, through the surface of the
tube. *Sphodros rufipes* from the eastern
USA looks very similar but has redder
legs and its buried tube has a section that
extends for 8–10 in/20–25 cm up the
trunk of an adjacent tree.
DISTRIBUTION Throughout much of Europe
southward to North Africa, and eastward
to western Asia.

Funnel-Web Tarantulas *Family Dipluridae*
Most members of this small, mainly tropical family can be recognized by the very long spinners. These jut out from the rear of the spider like two tails, and can be more than half as long as the abdomen. These spiders usually construct a broad sheet-web, which can attain 1 yd/1 m in diameter, although some species live in burrows in soft ground.

ANAME sp.

COMMON NAME Australian trap-door spider
DESCRIPTION Despite its common name, this species builds a burrow that lacks a door over the entrance. The 1 in-/25 mm-long male (illustrated) has special spurs towards the tips of his front legs (the left-hand spur is clearly visible in the illustration). These spurs are jammed against the female's fangs during the act of mating, preventing them from operating. At night the males wander in search of females' burrows. If provoked they rear up, open-fanged, in a defensive stance, ready to strike.
DISTRIBUTION Restricted to Western Australia, but similar species occur in other parts of Australia. These include the blackish-brown, stout-bodied Sydney funnel-web spider (*Atrax robustus*), renowned for its deadly bite.

Tarantulas or Bird-Eating Spiders *Family Theraphosidae*
This family includes the largest of all spiders, the goliath tarantula *Theraphosa blondi*, with its 10 in/25 cm legspan. Most tarantulas are very hairy and have a surprising ability to climb on smooth surfaces, aided by dense clusters of hairs on the tips of their legs. The eyes are small and packed closely together. Most species live in burrows in the ground but some live in trees. Prey as large as lizards, mice, and small birds may be taken. Tarantulas are mainly tropical, although some 30 species occur in the USA, with none in Europe.

BRACHYPELMA SMITHI

COMMON NAME Mexican red-knee tarantula
DESCRIPTION This is the most strikingly patterned of all the tarantulas, with its knees boldly marked in bright orange or red. The carapace is black, with a broad, brownish-red margin, and the abdomen is black or dark brown. The female can grow to 3 in/ 75 mm in length, but the male only reaches 2¼ in/56 mm. Like most tarantulas, the adult females are long-lived and spend most of their lives inside deep burrows. The status of this species in the wild has been threatened by the collection of huge numbers of adults for the exotic pet trade, as illustrated.
DISTRIBUTION Mexico.

APHONOPELMA CHALCODES

COMMON NAME Western desert tarantula
DESCRIPTION This is one of more than
20 members of this genus found in the
southwestern states of the USA. This
species is often very common in the deserts
of Arizona, and on certain nights large
numbers of males go on the move, searching
for females' burrows. The legs are dark and
the carapace is densely covered with sandy-
colored hairs, while the abdomen is more
sparsely clothed with russet hairs, allowing
a considerable amount of black to show
through. Males reach a length of 1¾ in/
44 mm, females 2¼ in/56 mm.

DISTRIBUTION Arizona, USA, and adjacent
areas of northern Mexico.

HARPACTIRA GIGAS

COMMON NAME Common baboon spider
DESCRIPTION Like most tarantulas, the
baboon spider is seldom encountered
outside its burrow. If threatened, it rears
up, leans backwards on its hind legs, holds
its front legs out at its sides and bares its
fangs, ready to sink them home. However,
the poison is neither potent nor abundant.
The long spinners can be seen projecting
from the rear of the abdomen, while the
carapace bears a spoke-like, radiating
pattern of brown lines on a black
background. Males can reach 2 in/50mm
in length, females ⅓ in/5 mm more.
DISTRIBUTION Cape region of South Africa,
northward to the Transvaal.

Suborder Araneomorphae

The great majority of spiders, both as species and as families, belong in this suborder. The most important difference between these spiders and the mygalomorphs lies in the side-to-side operation of the araneomorph fangs. The members of this suborder are considered to be more advanced and are capable of making many different kinds of silk. The sexual organs are often extremely complex.

Spitting Spiders *Family Scytodidae*
This family of six-eyed spiders can easily be recognized by the hump-backed appearance given by the extreme enlargement of the carapace into a conspicuous dome. This serves to accommodate the enlarged venom glands, which not only manufacture poison but also a special type of glue that is squirted from each fang, forming sticky threads that can immobilize prey up to ¾ in/9 mm away.

SCYTODES THORACICA

COMMON NAME Common spitting spider
DESCRIPTION The eyes of this inconspicuous and rather slow-moving, ¼ in/6 mm-long spider are arranged in three pairs. The domed carapace is obvious from the illustration, as are the yellowish legs ringed with black. The carapace and abdomen are heavily marked with black flecks. As in all spitting spiders, the female (illustrated) carries a ball of eggs in her jaws. There is no web, and the spiders can usually be found under rocks and stones, and in caves and buildings.
DISTRIBUTION Cosmopolitan.

Brown or **Recluse Spiders** *Family Loxoscelidae*
These rather small, brown, six-eyed spiders look harmless enough but are notorious for the severe symptoms resulting from their bites. They build rather delicate, scrappy sheet-webs of sticky silk in dark places such as rock crevices, barns, and houses. There are only a few species worldwide, mostly in the Americas.

LOXOSCELES RUFESCENS

COMMON NAME Violin spider
DESCRIPTION The long legs, held outwards to the sides, away from the rather slender brown abdomen, are typical of the group. The carapace bears a dark mark that has been likened to a violin, hence the common name. This ⁵/₁₆ in-/8 mm-long species builds a small sheet-web under stones, and in caves and buildings, but also roams actively in search of prey. It is not aggressive and is reluctant to bite humans. The brown recluse (*L. reclusa*) from the USA is much more venomous and prone to bite, which can cause severe ulceration. It is larger and plumper than the violin spider and has a shinier abdomen.
DISTRIBUTION Native to southern Europe and North Africa but introduced by man to a number of areas, including Japan, North America, Australia, and New Zealand, in most of which it is thriving and spreading.

Daddy Long-Legs Spiders *Family Pholcidae*

The extremely long, thin legs, with flexible ends, give these spiders a gangling look, which earns them their common name. Most species have eight eyes, among which the front-facing pair at the center are much smaller than the rest. A few have lost a pair of eyes and only have six, in two groups of three. More than 300 species are known worldwide.

PHOLCUS PHALANGIOIDES

COMMON NAME Common daddy long-legs spider or long-bodied cellar spider
DESCRIPTION This ½ in/13 mm-long spider

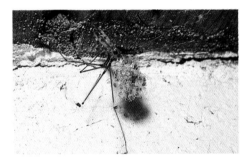

is usually seen hanging motionless for days on end in its large, tangled web in the corner of a room or outhouse. The pale abdomen bears a few dark marks along the top and there is a dark blotch on top of the carapace. The female uses her chelicerae for carrying a ball of eggs, loosely wrapped in a few strands of silk. This species is adept at killing other spiders, including large and powerful house spiders (*Tegenaria* spp.).
DISTRIBUTION Cosmopolitan, mainly in buildings, cellars, and caves.

PHYSOCYCLUS GLOBOSUS

COMMON NAME Short-bodied cellar spider
DESCRIPTION The dark abdomen of this ¼ in-/6 mm-long spider is rather short and plump, and from the side it is triangular in outline. On the face there are eight eyes crowded together on a dark prominence. A blackish band runs up the center of the carapace and is continued onwards for three-quarters of the length of the abdomen. The legs are brown. *P. californicus* from California has a dark, Y-shaped mark on top of the carapace and a pair of dark spots on top of the abdomen.
DISTRIBUTION Throughout the warmer parts

of the world, mainly in buildings, cellars, and caves.

Desert Bush Spiders *Family Diguetidae*
This is a small family of spiders with fewer than 20 species, found in the deserts of the USA, Mexico, and Argentina. They have six eyes in three groups and make characteristic webs in desert shrubs, such as creosote bushes and on prickly-pear cacti. These webs, with their pyramidal-shaped lines, are conspicuous from a considerable distance.

DIGUETIA CANITIES

COMMON NAME Desert bush spider
DESCRIPTION The brownish-orange cephalothorax is rather long and covered in a pelt of very short white hairs. The densely hairy abdomen is light brown and bears on its upper surface a darker, leaf-like shape (folium) bordered with white. The legs are mainly yellowish brown. The easiest way to recognize this ³/₈ in-/9 mm-long spider is by its web. This consists of a maze of threads connected to a domed sheet, surmounted by a tubular retreat richly decorated with plant remains and (in females) containing the egg-sacs. The web is often placed some 1–2 ft/ 30–60 cm off the ground in bushes.
DISTRIBUTION Restricted to the USA, from Oklahoma and Texas westward to California, in desert and semidesert.

Six-Eyed Spiders *Family Dysderidae*
This is a small family of spiders having six closely grouped eyes. Rather than two respiratory slits on the underside of the abdomen, there are four, and these are clearly visible. The members of this family are slow-moving, nocturnal hunters with large jaws.

DYSDERA CROCATA

COMMON NAME Woodlouse spider
DESCRIPTION The carapace is a rich reddish brown, while the egg-shaped abdomen is a light tan or gray. The rather thick legs are reddish orange, slightly lighter in tone than the carapace. The fangs are very long and project conspicuously forwards, contributing to an overall length of ⅝ in/16 mm in the females, and ⁷/₁₆ in/11 mm in the males. The fangs need to be large in order to penetrate the tough exterior of woodlice, which comprise the main prey of this spider. During the day the spider rests in a flattened silken retreat beneath a stone or log. The bite is painful but not dangerous and is not freely given.

DISTRIBUTION Native to Europe, but now found almost worldwide and common in North America.

Tube-Web Spiders *Family Segestriidae*
Formerly included in the previous family, these spiders now have a small family to themselves. The eyes, six in number, are again closely grouped, and the body is rather tubular, with the front three pairs of legs directed forward for grasping prey. The spiders spend their lives in tubes built in crevices in rocks, walls or tree bark, rushing out at night to seize insects which have blundered across the silken triplines radiating from the tube's entrance.

SEGESTRIA SENOCULATA

COMMON NAME Snake-back spider or leopard spider

DESCRIPTION A row of black spots down the middle of the abdomen, somewhat resembling the pattern of certain snakes, is a distinctive feature of this ³/₈ in-/9 mm-long spider. However, in some specimens (such as the one illustrated) the spots coalesce into an unindented dark band. The dark, shiny brown carapace is rather elongated and the legs are pale brown with a few darker brown rings. The habits of this common species correspond with those given above for the whole family. *Ariadna bicolor*, which occurs throughout the USA, has a similar outline but is a plain dark brown.

DISTRIBUTION Throughout most of Europe, then eastward through Asia to Japan.

Net-Casting Spiders *Family Deinopidae*

The members of this mainly tropical family are also called ogre-faced spiders, on account of two huge, headlight-style eyes staring menacingly from the rather small face (the six other eyes are very small). These huge eyes are essential aids to night vision. Prey is caught in a small net that the spider casts with considerable accuracy. Some species snare their flying insects in mid-air, while others thrust the net downwards to catch crawling insects on the ground beneath.

DEINOPSIS LONGIPES

COMMON NAME Net-casting spider

DESCRIPTION During the day this very slim ⅝ in-/16mm-long brown spider sits motionless, head-downwards, and is very difficult to distinguish from a twig. At night it constructs its prey-catching net of hackled-band silk. It holds the finished net in a collapsed state at the tips of its front legs and then expands it instantly to its full size while taking aim at a passing insect. *D. spinosa* from the southeastern USA is similar but slightly larger.

DISTRIBUTION Central America.

Feather-Footed Spiders *Family Uloboridae*
More than 200 species of uloborids have been described, mainly from the tropics. One of the main characteristics of this family is the absence of any poison glands. Another peculiarity is the possession of a cribellum, which is used for making an orb-web from hackled-band silk. In most species the front legs bear plumed feet, hence the common name.

MIAGRAMMOPES sp.

COMMON NAME Stick spider (new and undescribed species)
DESCRIPTION Stick spiders build single-line webs of very sticky silk, in which the spider's body forms part of the span. If an insect alights on the convenient perch offered by the single strand of silk, the spider lets go a short length of slack silk so that the line sags and entangles the insect. *M. mexicanus* from Texas is similar.
DISTRIBUTION Indonesia.

Pigmy Mesh-Spinners *Family Dictynidae*

This is the largest family of cribellate spiders, with over 500 species so far described worldwide. Most species are ¹⁄₁₀ in/3 mm long or less, and the family is mainly distributed in the temperate zones. The small and rather haphazard webs are constructed on leaves, twigs, and in crevices.

NIGMA PUELLA

COMMON NAME Leaf lace-weaver

DESCRIPTION The pale green, rather downy abdomen of the female (illustrated) is strikingly barred with light maroon and there is a large, maroon, rather diamond-shaped blotch near the front. The reddish-brown male is nearly as large as the female. This species lives in a horizontal and rather lacy web built close above the surface of a leaf. More than 150 species of dictynids occur in the USA, most of which are small brown spiders, although some species are quite attractive and are mostly white, yellow or red.

DISTRIBUTION Europe and North Africa.

Large Lace-Weavers *Family Amaurobiidae*

This family of some 350 species contains mainly brown or black spiders with a superficial resemblance to the funnel-weavers in the family Agelenidae. However, the clearly visible cribellum at the tip of the amaurobiid abdomen produces hackled-band silk and serves easily to distinguish the two groups.

AMAUROBIUS FENESTRALIS

COMMON NAME Window lace-weaver
DESCRIPTION The male of this common spider reaches a length of ⁵/₁₆ in/8 mm, while the female (illustrated) can grow up to ³/₈ in/ 9 mm and is considerably fatter. The legs and carapace are a glossy brown and the legs are decorated with a series of darker brown rings. The top of the abdomen bears a dark, wedge-shaped mark girdled by gold. The untidy webs are common around windowframes, under the eaves of houses, and any other places offering a handy crevice, such as the trunks of gnarled old trees. As in several other spiders, the mother's dead body is not wasted but will provide a funeral feast for her offspring.
DISTRIBUTION Europe and western Asia.

AMAUROBIUS FEROX

COMMON NAME Black lace-weaver
DESCRIPTION The plump black ⁵/₈ in-/16 mm-long females are usually found sitting beside their white egg-sacs in a nest underneath a log, rock, or plank of wood. The color of the body is a very dark brown and the abdomen bears a yellowish-brown pattern faintly resembling a human face. The male is slimmer than the female (illustrated) and shorter (⁷/₁₆ in/11 mm).
DISTRIBUTION Europe and North America; recently introduced to New Zealand.

Hole Spiders *Family Filistatidae*

This small family contains fewer than 50 species distributed around the warmer parts of the world. A dozen species occur in the USA. Their most distinctive feature is the rather elongated and pointed carapace, like the prow of a boat, with eight eyes closely grouped together on a small tubercle. There is no epigynum in the female.

KUKULCANIA ARIZONICA

COMMON NAME Arizona black hole spider
DESCRIPTION This black spider has a beautiful, lustrous, velvety sheen. The carapace is distinctly pointed and the male has longer legs than the female (illustrated). During the day the spider rests in a silken tube inside a hole or crevice. A number of silk lines radiate from the mouth of the tube, forming a web, and these are often conspicuous on the walls of buildings. The ½ in-/13 mm-long females live for several years.
DISTRIBUTION Arizona, USA.

Cobweb-Weavers or Comb-Footed Spiders *Family Theridiidae*

This is one of the largest families of spiders, with over 2,200 species worldwide. There are three claws on each leg and most species have eight eyes. The name "comb-footed" stems from the presence of a comb of bristles on the hind legs, which enables the spider to throw silk rapidly, and from a safe distance, over a struggling insect.

THERIDION MELANURUM

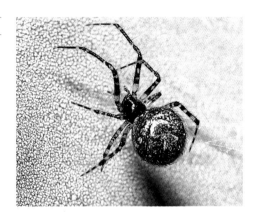

COMMON NAME Lesser glasshouse spider
DESCRIPTION As in all members of this family, the spider hangs upside-down in its rather scrappy scaffold-web. The carapace is dark brown, while the abdomen is black and gray. A jagged-edged band down the center of the abdomen is of equal width from front to rear. The light brown legs are dark ringed. The females can usually be found sitting beside their pale brown egg-sacs in the corner of a room or outhouse.
DISTRIBUTION Mainly in houses in Europe and North America.

ACHAEARANEA TEPIDARIORUM

COMMON NAME Glasshouse spider
DESCRIPTION The sides of the globular blackish abdomen are streaked with brown

or gray, while down its center there is a row of pale chevrons. The web is usually constructed in the corner of a room. The outer threads of the web are taut and heavily beaded with glue, which entraps insects much larger than the spider. The pear-shaped egg-sac is suspended in the web, usually with the female on guard beside it. The ¼ in/6 mm long females live more than one year; the males are only ⅛ in/3 mm long.
DISTRIBUTION Cosmopolitan, in houses.

ENOPLOGNATHA OVATA

COMMON NAME Red and white cobweb-weaver

DESCRIPTION The female (illustrated) reaches ¼ in/6 mm in length, and has pale, almost transparent legs and a shiny, brownish-green carapace. The white or cream abdomen can be marked in three different ways: first with a row of dark spots, second with a broad red band, or third with two red stripes. The females are giant-killers, and will fearlessly tackle large stinging insects, such as the bumble-bee in the illustration. The female stands guard over her grayish-blue egg-sac inside a curled leaf fastened together with silk.

DISTRIBUTION Native to Europe, but introduced to North America where it is

now a common and familiar spider.

ARGYRODES FLAVESCENS

COMMON NAME Red and silver dewdrop spider

DESCRIPTION The reddish-brown, very shiny abdomen bears a number of silver spots and, as in most dewdrop spiders, it is high and domed, with an almost triangular outline. Most members of this genus are tiny (⅛ in/3 mm long or so) and live in the webs of other larger spiders (usually orb-web spiders such as *Nephila* and *Argiope*), feeding on the prey caught by the host spider. *A. trigonum*, a common species in the USA, looks rather similar, but is more yellowish and lacks the silver spots.

DISTRIBUTION Sri Lanka and Burma eastward through Indonesia.

STEATODA GROSSA

COMMON NAME Cellar spider
DESCRIPTION The male (illustrated) reaches ¼ in/6 mm in length. The legs are a pale, almost transparent brown, the carapace a glossy, deep brown (almost black), and the abdomen black with several greenish-brown marks, on which there is a peppering of white spots. The larger female (up to ⅝ in/16 mm long) has a much rounder, glossy black abdomen, with no visible markings (except sometimes a pale semicircle and three spots), and a black carapace and legs. The female suspends her fluffy white egg-sacs in her web, which is usually found in cellars, caves or hollow trees. She guards her egg-sacs until the babies emerge, after which they remain

in her web for some time, sharing her food.
DISTRIBUTION Cosmopolitan.

LATRODECTUS MACTANS

COMMON NAME Black widow spider
DESCRIPTION The red "hourglass" marking on the underside of the shiny black abdomen is the unmistakable trademark of the black widow spider. The round-bodied females

reach a length of 9/16 in/14 mm but the much paler males only reach 3/16 in/5 mm. The very scrappy web is often found in and around buildings – the web in the illustration was underneath a table in a house. The bite is highly venomous, but seldom lethal in man, and the black widow is a nervous spider which will drop from its web at the slightest disturbance. In the almost identical species, *L. hesperus*, enemies such as mice are repelled by having their faces smeared with sticky silk that the spider deploys in a kind of defensive net.
DISTRIBUTION Many of the warmer parts of the world, including southern USA.

CHRYSSO sp.

COMMON NAME Triangular spine-leg spider
DESCRIPTION This is a very strange-looking spider. The high-peaked abdomen is triangular in profile and very flattened. The unusually long legs are fringed with numerous, stiff black spines. The untidy web is hung beneath the foliage of trees and bushes, and the $^{5}/_{16}$ in-/8 mm-long female hangs upside-down, often beside her tan-colored, bell-shaped egg-sac. The species illustrated is presently undescribed. *C. pulcherrima* is a cosmotropical species which is similar but less spiny.
DISTRIBUTION Indonesia.

Dwarf Spiders or Money Spiders *Family Linyphiidae*
This is a large family of mostly very small spiders which are more common in the cooler parts of the world, such as Europe and North America, than in the tropics. The males of many species are provided with stridulatory ridges on the chelicerae which are used to produce vibrations of the web during courtship. In addition, there are many instances where the head of the male is embellished with bizarre knobs and turrets which play a part in the mating process. The web is usually a small hammock. Dwarf spiders are the most abundant of all spiders and their populations can reach 2¼ million per acre/5½ million per hectare in a suitable habitat, such as a grassy field.

NERIENE PELTATA

COMMON NAME Platform-web spider
DESCRIPTION This is one of several species of black (or brown) and white dwarf spiders which can be seen hanging beneath their hammock-webs during the day. The abdomen of this ³/₁₆ in-/5 mm-long species bears a narrow white central stripe, bordered on either side by a wavy-edged brown band, flanked by white. The carapace is light brown with a triangular blackish central band. The web is mainly built in bushes and among the lower branches of trees.
DISTRIBUTION Europe, eastward to Japan, and in the USA.

DRAPETISCA SOCIALIS

COMMON NAME Invisible spider
DESCRIPTION This ⅛ in-/3 mm-long spider
spends the day sitting motionless on the
trunk of a tree. The mottled, brown and
white abdomen, and banded brown and
black legs blend in very well with the colors
and textures of tree bark. Usually the spider
is perched on top of a fine web that hugs
the bark. *D. alteranda* from the USA is very
similar and occurs on tree trunks from
New England westward to Wisconsin.
DISTRIBUTION Throughout Europe and most
of temperate Asia.

MICROLINYPHIA PUSILLA

COMMON NAME Dainty platform spider
DESCRIPTION The shiny black male
(illustrated) reaches a length of ³⁄₁₆ in/5 mm
and has a tubular abdomen with two white
spots on the top, near the front. The slightly
larger female is completely different and
looks more like *Neriene peltata* (page 31),
with a plump silvery abdomen bearing a
broad, black, leaf-like mark. The small
hammock-web is built among grass and in
low vegetation, often in damp places.
DISTRIBUTION Europe, temperate Asia, and
North America.

Funnel-Weavers or Grass Spiders *Family Agelenidae*
This is a large family containing more than 1,000 species of mainly brownish spiders, varying in length from $\frac{1}{16}$ in/2 mm to 1 in/25 mm, and usually having long and conspicuously hairy legs. They construct a broad, flat sheet-web of non-sticky silk leading down into a tubular retreat. Intersecting knock-down lines deployed above the sheet bring flying insects crashing down into the web.

AGELENOPSIS APERTA

COMMON NAME Desert grass spider
DESCRIPTION There are numerous species of grass spiders found throughout the USA but these can only be reliably distinguished from one another by an expert. The males average $\frac{5}{8}$ in/16 mm in length, the females $\frac{3}{4}$ in/19 mm. There are eight eyes, arranged in two rows. Most species have a pair of dark stripes on either side of the carapace. The spider illustrated is one of the grayer species; most of the others are brown. There is usually some kind of pale, longitudinal striping on the abdomen, from which the spinners protrude conspicuously. The spider lurks at the entrance to its tubular retreat, ready to rush out and pounce on arriving insects. The females lay eggs in autumn, then die. *Agelena labyrinthica*, from Europe, is similar but brown.

DISTRIBUTION Numerous similar species are found throughout North America.

TEGENARIA DUELLICA

COMMON NAME Cobweb spider
DESCRIPTION This is the spider that is most likely to cause panic as it sprints rapidly across the carpet on its long, hairy legs, or crouches menacingly in the bathtub. It is, of course, perfectly harmless but is responsible for eliciting fear and loathing in countless

people. The ½ in-/13 mm-long males are often the culprits, as they abandon their large sheet-webs in the corner of a room or outhouse and wander off in search of females. These Romeos usually move in with the slightly larger females and they will live together amicably for a considerable time before the male finally dies. His corpse provides his mate with a substantial meal that will make a significant contribution towards the development of her eggs, in effect converting father into father's offspring. This spider, along with several of the other large species of *Tegenaria*, can endure long periods of fasting and the females may live for several years.
DISTRIBUTION Europe and North America.

TEGENARIA AGRESTIS

COMMON NAME Yard spider
DESCRIPTION This European spider has been introduced to the northwestern states of the USA, where it has become common around houses. Its large, untidy-looking sheet-web is usually found in backyards. It is smaller and rather grayer than the cobweb spider, the female (illustrated) reaching a length of ⁹/₁₆ in/14 mm but the male only ³/₈ in/9 mm. The spherical egg-sacs are placed in the web structure and are camouflaged with bits of rotting wood, particles of soil or insect skeletons. Numerous instances of this species biting humans have been reported but the bite is not serious.

DISTRIBUTION Europe and northwestern North America.

Long-Jawed Orb-Weavers *Family Tetragnathidae*

The orb-webs constructed by these usually elongated spiders are seldom vertical but are usually inclined at an angle and sometimes horizontal. With the exception of *Leucauge* the females lack an epigynum. The jaws in both sexes are large and prominent, and the males of some species have special prongs for jamming open the female's jaws during mating. Members of the genus *Pachygnatha* have more rotund bodies and do not construct webs as adults. Within this family are spiders of the genus *Nephila*, the largest of all web-building spiders. The incredibly strong yellow webs of *Nephila clavipes* in the USA have been recorded spanning gaps as much as 60 feet across.

TETRAGNATHA EXTENSA

COMMON NAME Common long-jawed orb-weaver

DESCRIPTION The elongated body and legs are typical for the genus. This ³⁄₈ in-/9 mm-long species is most likely to be found near water. During the day the spider sits in the center of the web or on a nearby leaf or grass stem, with the long front and back legs held out fore and aft. They often lie hidden from their predators and in wait for possible prey items. The egg-sacs are placed on a leaf and are covered in tufts of grayish-green silk. These spiders build impressive webs that are capable of trapping larger warm-blooded animals, such as humming birds and bats, although the normal prey is much smaller.

DISTRIBUTION Europe and most of temperate Asia, and in North America.

METELLINA SEGMENTATA

COMMON NAME Common orb-weaver
DESCRIPTION This is probably the commonest orb-weaver in Europe. The $^5/_{16}$ in-/8 mm-long female (illustrated) is pale yellowish brown and there is a darker pattern resembling a leaf or fir tree on the abdomen. The male has longer legs and is a more rusty brown, with a smaller and slimmer abdomen. During the mating season the males take up residence in the female's web but wait until an insect is trapped before commencing courtship. They have to entice the female on to a special mating thread before consummation can take place.
DISTRIBUTION Europe, eastward to Asia, and in Canada.

LEUCAUGE NIGROVITTATA

COMMON NAME Black-striped orchard spider
DESCRIPTION Orchard spiders can be easily recognized wherever they are seen throughout the world, as they all have a very similar and highly characteristic appearance. This ½ in-/13 mm-long species has a silver stripe down the center of the abdomen, dissected lengthwise by three black lines; the flanks are greenish yellow. The webs are inclined at 45° to the vertical and are often grouped together.

Two species of orchard spiders occur in the eastern USA. In the Mabel orchard spider (*L. mabelae*) the yellowish abdomen bears eight, evenly spaced silver bands, and three orange spots. In the venusta orchard spider (*L. venusta*) the top of the abdomen is silver, striped with black, the sides are yellow, and there are two red spots on the underside. Both species build an almost horizontal web in the lower branches of shrubs and trees.
DISTRIBUTION Indonesia.

NEPHILA CLAVIPES

COMMON NAME Golden-silk spider or golden orb-weaver

DESCRIPTION The females are the giants among orb-weavers and can reach 1⅜ in/ 34 mm in length. The front two pairs of legs, along with the rearmost pair, all bear tufts of dark hairs. The sausage-like abdomen is decorated with white or gold spots and flecks, while the carapace is silver. The brown males are only ⅜ in/9 mm long. The huge webs are made of tough yellow silk, and often carry uninvited dewdrop spiders (*Argyrodes* spp.) klepto-parasites. In the USA the adults die in winter but in the tropics they are present throughout the year.

DISTRIBUTION Southeastern states of the USA, southward to northern Argentina.

NEPHILA SENEGALENSIS

COMMON NAME African golden orb-weaver

DESCRIPTION The top of the female's abdomen has a broad black or silver gray, rather jagged-edged band in which are set four pairs of pale cream spots. These are often united to form discrete bands or else form a dumb-bell shape. The sides of the abdomen are also cream, and the carapace is blackish silver. The legs are banded in black and dark brown. The juveniles are much more silvery, with numerous whitish spots, and are often seen sitting below a dense stabilimentum of thicker silk, adorned with numerous insect remains. Adult females may be 1⅜ in/34 mm in length, but the brown males only reach ¼ in/6 mm and are usually present in the females' webs.

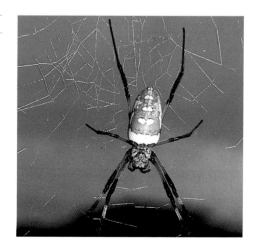

DISTRIBUTION Most of sub-Saharan Africa, in desert, savannah, and forest.

Orb-Weavers *Family Araneidae*

This is a large and varied family, with over 3,500 species around the world. Although the perfect orb-web is perhaps a typical feature of the family, there are huge variations in its structure, and some species have dispensed with a web entirely and reverted to a life of ambush. Orb-weavers have tiny eyes that play only a minor role in daily life. The sense of touch is far more important, and these spiders monitor events in their webs through their vibration-sensitive legs and feet.

ARANEUS DIADEMATUS

COMMON NAME Garden or cross spider

DESCRIPTION The color is very variable and can be pale fawn, deep rusty brown, bright orange, brownish-gray and every tone in between. Whatever the ground color, in the female (illustrated) there is usually an unmistakable cross on top of the abdomen, formed by a series of white blotches. The spider sits in the center of its web or on a plant nearby. Most prey is bitten first, then wrapped. Females attain a length of ⅝ in/ 16 mm and become very plump when they are full of eggs; the slimmer male does not exceed ⅜ in/9 mm.

DISTRIBUTION Europe, eastward to Japan. Introduced to North America and now very common in many areas, especially in the east.

ARANEUS QUADRATUS

COMMON NAME Four-spot orb-weaver
DESCRIPTION The female has a rotund abdomen which is usually tan colored but can be rusty brown or greenish yellow. There are always four prominent white spots on the front half of the abdomen, plus a number of small white dots and squiggles. The legs are attractively banded. The spider rarely sits in its web, but can usually be located sitting head-downward in a lair made by fastening together a number of plant stems with a dense mesh of silk. The shamrock spider (*A. trifolium*), common throughout North America, more or less combines the patterns of the four-spot orb-weaver and the garden spider (*A. diadematus*), with the four spots of the former and the vertical section of the cross of the latter.

However, it has a white carapace with a central black stripe, and another black stripe on each side, a distinctive character absent in the other two species.
DISTRIBUTION Throughout Europe and much of Asia.

ARANEUS ILLAUDATUS

COMMON NAME Texas orb-weaver
DESCRIPTION This is a very large, grayish-white or sometimes even pinkish-white species in which the female (illustrated) can reach 1 in/25 mm in length. The top of the abdomen (near the front) bears two blackish-brown markings, which are more or less triangular and have very indented, rather ragged edges. Each of these darker markings contains a white dot, and there are some more, rather leaf-like markings towards the rear of the abdomen. The whole spider is extremely hairy. The male is a pigmy of only ⅜ in/9 mm.
DISTRIBUTION Southeastern Arizona to western Texas.

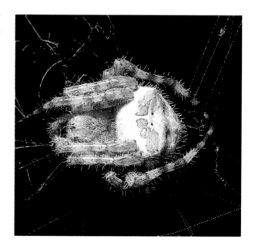

ARANEUS MARMOREUS

COMMON NAME Marbled orb-weaver

DESCRIPTION There are two forms of this species. In the one illustrated (var. *pyramidatus*) the female's plump abdomen varies from pale whitish-cream to a rich golden yellow. Towards the rear there is a dark brown, pyramid-shaped blotch. In the other form the abdomen is a much darker shade of yellowish orange and bears a definite brown or blackish pattern of markings with zigzag edges, giving a marbled effect. The two forms are seldom found together and var. *pyramidatus* only occurs in the European populations. The female reaches a length of ⁹/₁₆ in/14 mm and the male ³/₈ in/9 mm. The web is placed among low vegetation in open areas in woodland, and on heaths.

DISTRIBUTION Europe, eastward over most of Asia, and in most of North America but absent from the southwestern USA.

ACULEPEIRA CEROPEGIA

COMMON NAME Mountain orb-weaver
DESCRIPTION This is one of Europe's most attractive orb-weavers. The abdomen is rather egg-shaped, being pointed at both front and rear, and bears a handsome cream pattern resembling a rather slim fir tree. The abdomen overhangs rather at the back, so that the spinners are no longer terminal. The pale brown legs are heavily banded with numerous blackish-brown rings. The ½ in-/13 mm-long female (illustrated) can usually be found sitting on top of a platform of dense silk to one side of the web, which is attached to low vegetation near the ground.

DISTRIBUTION Much of Europe (but excluding the British Isles) eastward through Asia, and in North Africa.

NEOSCONA ADIANTA

COMMON NAME Bordered orb-weaver
DESCRIPTION This attractive spider is rather like a miniature version of the mountain orb-weaver. The ground color of the abdomen varies from a pale grayish brown to quite a dark shade of rusty red. A series of prominent white or cream triangles with black outer borders (these borders are sometimes only partial) runs down the center of the abdomen. The triangles towards the front are much larger than those at the rear. The ⅜ in-/9 mm-long female (illustrated) sits in full view beside the web, on some plant, often the fruiting head of an umbellifer such as wild carrot, on top of a small pad of silk.
DISTRIBUTION Local in the British Isles, and mainly found near the coast in the south. Throughout Europe and Asia to Japan, and in North Africa.

NUCTENEA CORNUTA

COMMON NAME Bankside orb-weaver or furrow spider

DESCRIPTION The carapace of this common spider is brown to gray, and densely covered in short white hairs. The ground color of the abdomen varies from silvery gray or white to cream, light brown or bright rusty red. On top there is a blackish, leaf-like pattern which varies somewhat in intensity from one individual to another. The female (illustrated) reaches a length of $7/16$ in/11 mm, and at $5/16$ in/8 mm the male is only slightly smaller. He generally takes up lodgings in the lair of an immature female before she makes her final molt.

DISTRIBUTION Europe, eastward through Asia, and in most of North America.

ARANIELLA CUCURBITINA

COMMON NAME Green orb-weaver

DESCRIPTION The carapace of the $1/4$ in-/6 mm-long female is glossy brown, the legs a slightly more greenish brown. The top of the abdomen can be yellow or yellowish green and has four conspicuous pits arranged more or less in a square. The underside is normally bright green, and there is a bright red mark just behind the spinners. The web is rather small and is often constructed across a large, curled leaf or between a number of smaller leaves. The females are often found standing guard over their egg-sacs, which are covered in coarse yellow silk.

DISTRIBUTION Europe, eastward to Japan, and in North Africa.

ARGIOPE APPENSA

COMMON NAME Great golden argiope

DESCRIPTION The female (illustrated) can often reach a length of 1¼ in/32 mm. The top of the abdomen is a bright lemon yellow, with a pair of large black dots, and two smaller ones. The legs are strongly banded in black and gray. The webs are often built in gardens, or even under the eaves of houses, where several may occur together.

DISTRIBUTION Southeast Asia.

ARGIOPE ARGENTATA

COMMON NAME Silver argiope
DESCRIPTION In this handsome spider the carapace and the front half of the abdomen are bright silver. The rear half of the abdomen is brightly decorated with silver markings on a yellow or orange background, and there are five prominent lobes around the rear margin. The shape of the abdomen is very characteristic, as it is narrow at the front and then flares out towards the rear. The legs are banded in black and yellow.

The female (illustrated) can reach a length of ¹¹/₁₆ in/17 mm, but as in all *Argiope* spp., the male is a midget and only reaches ³/₁₆ in/5 mm. He is also much more lightly built than the female. The web is built near the ground among grasses and shrubs.
DISTRIBUTION From southern USA (southern Florida westward to southern California) to northern Argentina.

ARGIOPE BRUENNICHI

COMMON NAME Bruennichi's argiope or
wasp spider
DESCRIPTION The plump, rather egg-shaped
abdomen of the 1 in-/25 mm-long female
(illustrated) is decorated with a series of
closely spaced, black, yellow and silver
bands, while the carapace is plain silver,
and the legs are banded light and dark.
The web is built among grass near the
ground and mostly catches jumping insects
such as grasshoppers. The female often
starts to wrap the tiny brown male in silk
while they are mating.

DISTRIBUTION Europe and temperate Asia,
eastward to Japan.

abdomen is heavily lobed around the rear
margin, and the apical lobe is often long
and pointed, so that the abdomen rather
resembles a pixie's hat. The basic color is
silver, with a number of small black pits
and two red spots. The female (illustrated)
reaches a length of 1 in/25 mm but the
males are tiny at only ¼ in/6 mm. As in all
members of the genus, the center of the web
is usually decorated with a stabilimentum
of dense white zigzag silk. This can be
constructed in the form of a St. Andrew's
Cross (often with one or more of the four
segments missing) or a vertical line, as in
the illustration. In all cases there is a gap
in the middle where the spider sits.
DISTRIBUTION Southern Europe, eastward
to Asia, and in the whole of Africa.

ARGIOPE LOBATA

COMMON NAME Lobed argiope
DESCRIPTION The rather furrowed, shiny

ARGIOPE TRIFASCIATA

COMMON NAME Banded argiope
DESCRIPTION The rather hairy, pointed abdomen of the 1 in-/25 mm-long female (illustrated) bears a series of alternating cream, black, and silver bands. These all have more or less straight edges. The legs are spotted, and the carapace silvery. As in all members of this genus, the tiny males often spend many days in residence in the female's web. The web is suspended among tall grass and herbaceous plants.
DISTRIBUTION Throughout the warmer parts

of the world. Rare in southern Europe, but common throughout the USA.

HERENNIA ORNATISSIMA

COMMON NAME Ornate orb-weaver
DESCRIPTION The dull gray, flattened abdomen of the female (illustrated) is distinctly crenellated around the rear margin and heavily dotted with reddish spots. In the center of the abdomen there are two blackish-red spots situated in deep pits. The carapace is bluish gray with a pale orange margin. The large web is built just above, and parallel to, the surface of a tree trunk and can only be seen from certain angles in favorable lighting conditions. The female sits on a cup of dense silk near the web's hub, where her coloration makes her very difficult to see. The shiny brown males are minute and are often resident on the edge of the female's web.
DISTRIBUTION From India to New Guinea.

CYRTOPHORA HIRTA

COMMON NAME Hairy tent-spider
DESCRIPTION The ground color of the $^9/_{16}$ in-
/14 mm-long female (illustrated) is silvery
white. The top of the abdomen is decorated
with a complex black pattern, and there is a
yellow and black pattern on its sides. The
spinners are situated on a tubercle, which
gives the rear of the abdomen a distinctively
prow-shaped profile from the side. The legs
are banded light and dark. The spider hangs
upside-down beneath a horizontal orb-
web, which is slung beneath a scaffold of
intersecting lines. These interrupt flying
insects and knock them down into the web.
None of the silk is sticky, which is rare for
an araneid. Several webs occur together.
DISTRIBUTION Australia.

ZYGIELLA X-NOTATA

COMMON NAME Missing sector orb-weaver
DESCRIPTION The webs of this common spider are often abundant and conspicuous on and around houses, especially on window frames, and beneath roof overhangs. The webs are easily recognized by the absence of one section of the orb, near the top. During the day the spider hides in a silken lair nearby. The smooth, silvery abdomen is marked with a darker leaf-like pattern and the legs are heavily banded. Females reach a length of $^5/_{16}$ in/8 mm while the males are slightly smaller but look very similar.

DISTRIBUTION Common in Europe, much of temperate Asia, and the whole of North America.

CYCLOSA CONICA

COMMON NAME Conical orb-weaver
DESCRIPTION The color of the abdomen varies from gray with black markings to almost pure black. The abdomen is distinctly, but bluntly pointed, at the rear end and slopes forward over the carapace at the front. The females can reach $^1/_5$ in/5 mm in length; the males are slightly smaller. The spider usually sits at the hub of its surprisingly large web. Above and below the hub there is usually a vertical string of prey remains incorporated into a dense silk stabilimentum. In *C. turbinata* from the USA the tip of the abdomen is extended into a longer and more acute point.
DISTRIBUTION Europe, Asia, and North America.

GASTERACANTHA ARCUATA

COMMON NAME Long-horned orb-weaver
DESCRIPTION The ³/₈ in-/9 mm-long female (illustrated) of this amazing spider can be mistaken for no other. From the top corners of the almost triangular, bright orange abdomen protrude two long, black, curved "horns." Two thinner and shorter spikes jut up from the rear margin of the abdomen, between the major horns. It is believed that the projections make it difficult for birds to grasp hold of them, though no one is absolutely sure. The legs are black and the rather inconspicuous carapace is dark brown. The tiny males are hornless.
DISTRIBUTION Asia, from India to Indonesia, often in gardens.

GASTERACANTHA HASSELTI

COMMON NAME Hasselt's spiny spider
DESCRIPTION In the $^5/_{16}$ in-/8 mm-long female (illustrated) the top of the almost triangular abdomen is glossy bright orange, with a diverging row of six black spots on either side of the center line. The top corners are drawn out into two thick black spikes, which are more or less straight and have pointed tips. Two smaller black spikes protrude from the rear margin of the abdomen and two more from the sides, midway between the large spikes and the carapace, which is densely covered in short white hairs.
DISTRIBUTION India to Indonesia.

GASTERACANTHA STURII

COMMON NAME Blunt-spined kite spider
DESCRIPTION The very broad, kite-shaped abdomen is glossy yellow, crossed by two black lines. The top corners of the abdomen are armed with two stout, black, and rather hairy spines. These are blunt-ended save for a tiny thorn at the tips. Two very short spines project from the front part of the abdomen and two more from the rear. This stur's female kite spider sits on a leaf in the rain forest margin.
DISTRIBUTION Southeast Asia.

GASTERACANTHA FALCICORNIS

COMMON NAME Horned orb-weaver
DESCRIPTION The very hard, bright red, glossy abdomen of the female (illustrated) is deeply punctured with black pits. Two long curved horns and four short straight ones project from the abdomen. Sometimes a white or yellow band runs across the abdomen. This is quite a good example of a typical araneid orb-web although it is damaged with use.
DISTRIBUTION Eastern and southern Africa.

GASTERACANTHA CANCRIFORMIS

COMMON NAME Crablike spiny orb-weaver
DESCRIPTION The ½ in-/13 mm-wide abdomen of the female (illustrated) is subject to considerable color variation. The ground color can be white, cream, yellow or pale orange, and the six spurs can be black or red. Some forms are liberally speckled with black or brown spots, while others are virtually unmarked. The carapace is dark brown. The spider normally hangs from the underside of a sloping web, built near ground level. This female was found in a Mexican rain forest.
DISTRIBUTION USA, from North Carolina to Florida, and across to California; Central America.

GASTERACANTHA MINAX

COMMON NAME Christmas spider
DESCRIPTION In the female (illustrated) the ground color of the abdomen is black, with a white or yellow pattern. Six stumpy spurs circle the abdomen, giving it a star-shaped appearance. The width of the abdomen is about $7/16$ in/11 mm. The webs are constructed on shrubs and often occur in vast numbers over hundreds of acres.
DISTRIBUTION Over most of Australia.

MICRATHENA GRACILIS

COMMON NAME Lumpy thorn spider
DESCRIPTION The color of the abdomen can be white, pale yellow, yellowish brown or even black, while it can be virtually unmarked or spotted with brown. The $3/8$ in-/9 mm-long female (illustrated) has ten sharp spines. The spinners are set well forward on a deep tubercle, so that in profile the abdomen is almost triangular. The shiny carapace is elongate and the eyes are tiny. The male's spineless abdomen is long and almost parallel sided.
DISTRIBUTION USA, east of the Rockies; Central America.

MICRATHENA SAGITTATA

COMMON NAME Arrow-shaped thorn spider
DESCRIPTION The top of the abdomen is bright yellow, with a number of black puncture marks. Two long, deep-red spines protrude from the top corners of the abdomen and there are two pairs of shorter spines at even intervals along the sides. The yellow underside (illustrated, in a female) is patterned with red and black, and the spinners protrude downward on a distinct tubercle. Females are ³/₈ in/9 mm long, males only ³/₁₆ in/5 mm.
DISTRIBUTION Eastern USA and Central America.

MECYNOGEA LEMNISCATA

COMMON NAME Basilica spider
DESCRIPTION In its shape and color this ³/₈ in-/9 mm-long spider is rather similar to the *Leucauge* spp. orchard spiders. However, in the basilica spider the abdomen has a conspicuous hump on either side, near the base. The yellow carapace has a narrow black line down the middle and a wider line along each margin. The olive-green abdomen sports a black and brown, leaf-like pattern, bordered with white, and there is a wavy white line down each side. The snare consists of an orb-web arranged as a horizontal dome, set between upper and lower scaffolds of intersecting lines. The spider sits head-downward and, in the fall, the females will be hanging beneath a string of egg-sacs, as seen in the illustration.
DISTRIBUTION Most of the USA, but excluding the west-coast states.

Wolf Spiders *Family Lycosidae*

This is a large family of mainly brown or gray spiders, with more than 3,000 species worldwide. The majority are active hunters, or sit on leaves and pounce on insects that walk or fly past. Most wolf spiders live on the ground, and many do not have any kind of permanent home, although others live in silk-lined burrows, emerging after dark to hunt. Their vision is good, aided by two large eyes in the center of the face, flanked by two smaller eyes. A further row of four small eyes lies beneath them.

TROCHOSA TERRICOLA

COMMON NAME Earth-chaser
DESCRIPTION The legs are light brown, and the abdomen and carapace are a slightly darker brown, with few obvious markings. The ½ in-/13 mm-long female (illustrated) spends the day in a silken cell beneath a stone or log, emerging at night to hunt.
DISTRIBUTION Europe, much of Asia, and northern USA.

PARDOSA AMENTATA

COMMON NAME Spotted wolf spider
DESCRIPTION Dozens of similar-looking species of *Pardosa* wolf spiders are found in all kinds of habitats in Europe and North America (more than 50 species in the USA). As in all of them, the ⁵/₁₆ in-/8 mm-long female of this species carries her egg-sac attached to her spinners. The babies ride around on their mother's back for the first few days. The males are only slightly smaller than the females and employ a system of semaphore-signaling with the legs and palps during courtship.

DISTRIBUTION Europe and much of temperate Asia.

LYCOSA CAROLINENSIS

COMMON NAME Carolina wolf spider
DESCRIPTION This is the largest of a multitude of wolf spiders found in the USA. The females can reach a length of 1³/₈ in/34 mm, the males ³/₄ in/19 mm. The carapace is rather broad and can be quite dark, as in the specimen illustrated, with only a narrow pale central band. In some individuals the central band is broader and lighter and there are similar bands along both sides of the carapace. The abdomen varies from light brown to very dark brown, usually with a blackish-brown central band with either straight or jagged margins. The females spend much of their lives in a burrow in the ground but emerge at night to hunt, often carrying their large white egg-sacs attached to the spinners.
DISTRIBUTION Throughout the USA and much of Canada.

PIRATA PIRATICUS

COMMON NAME Common pirate spider
DESCRIPTION Members of this genus can be recognized by the pale V- or Y-shaped mark on top of the carapace. There are numerous similar-looking species that live beside freshwater in Europe and the USA. This $5/16$ in-/8 mm-long species has a light brown, velvety abdomen with a distinct sheen. A pair of pale lines runs from the front of the abdomen to the rear, gradually converging so that they eventually touch to form an elongated V. There is a line of small white spots toward the rear of the abdomen, whose sides are densely clothed in short, white hairs. The female (illustrated) carries her spherical white egg-sac attached to her

spinners and runs with it across the surface of the water. If alarmed, she will disappear beneath the surface for several minutes.
DISTRIBUTION Europe, eastward through Asia to Japan, and in North America.

ARCTOSA PERITA

COMMON NAME Sand-runner
DESCRIPTION The members of this genus are characterized by a rather flattened carapace, with the posterior eyes more or less mounted on top so that they look upwards. The carapace of the female (illustrated) varies from pale brown to black, the abdomen bears four large pale spots against a darker background, and the legs are ringed dark and light. This generally mottled effect gives excellent camouflage against a sandy background. The spiders spend part of their time in a burrow in the sand but also emerge to hunt in daytime. There are around a dozen *Arctosa* spp. in the USA, of which *A. littoralis* is very similar to the sand-runner, being found throughout the USA.

DISTRIBUTION Europe and much of Asia.

Nursery-Web Spiders *Family Pisauridae*

These are similar to the wolf spiders but the eyes are more or less all the same size. The female also carries her egg-sac, as in wolf spiders, but holds it beneath the front of her body, suspended from her fangs and pedipalps. The females make silken nurseries for their young. About 500 species occur worldwide.

PISAURA MIRABILIS

COMMON NAME Wedding-present spider
DESCRIPTION Males and females are of approximately equal size, averaging ⁹/₁₆ in /14 mm in length, but the male is of a slightly lighter build and has longer front legs. The ground color varies greatly and can be gray, light tan, chocolate brown or anything in between. Down the center of the carapace there is a white line, flanked by a dark band. The sides of the abdomen usually bear a dark, wavy-edged band, while the top usually has a series of faint chevrons. However, in some specimens all markings are virtually absent. The unique feature of this spider is the male's habit of presenting the female with a "wedding present" consisting of a fly densely wrapped in white silk. The female can often be seen trundling around with her large, spherical white egg-sac. Just before the young emerge, she fastens a few leaves or stems together with silk, places the egg-sac within, and then stands on guard. The American nursery-web spider (*Pisaurina mira*) looks similar, is the same size, and has similar nesting habits but does not present a nuptial gift. It is common throughout most of the USA east of the Rockies.
DISTRIBUTION Europe, eastward through most of temperate Asia, and in North Africa.

DOLOMEDES FIMBRIATUS

COMMON NAME European fishing spider

DESCRIPTION This dark brown spider is easily recognized by the white band that runs along both sides of the body. The female (illustrated) can reach a length of ¾ in/19 mm, the male only ½ in/13 mm. They sit on water, waiting to detect the ripples from an insect or fish. The females carry their large, grayish-green egg-sacs in their chelicerae. The six-spotted fishing spider (*D. triton*) from the USA looks very similar and has the same habits. It is common in swamps east of the Rockies.

DISTRIBUTION Europe and much of Asia.

Lynx Spiders *Family Oxyopidae*
These agile spiders can be recognized by the slim, tapering abdomen which ends in a point, long, heavily spined legs, and unique eye arrangement, with six larger eyes forming a hexagon and two smaller eyes below. The females guard their egg-sacs. Most species are found in the tropics, with only some 15 in the USA and even fewer in Europe.

OXYOPES SCHENKELI

COMMON NAME Bridal-veil lynx spider
DESCRIPTION The pattern of this ³/₈ in-/9 mm-long spider is so variable that it is impossible to describe in words, so perhaps likening it to a Persian carpet will adequately sum it up. It can leap into the air to capture a passing butterfly. Prior to mating, the female hangs in mid-air, suspended from her dragline, while the male wraps her in a silken bridal veil. This is the only lynx spider known to exhibit such behavior.

DISTRIBUTION Tropical Africa.

PEUCETIA VIRIDANS

COMMON NAME American green lynx spider
DESCRIPTION The normal color of this beautiful ³/₄ in-/19 mm-long spider is bright green but some individuals, especially in the western states of the USA, are yellowish or brownish. The top of the abdomen is usually decorated with a double row of red chevrons but these may be very faint. The closely spaced eyes are situated within a pale brown rectangle and the legs are very spiny. The females fix their knobbly, straw-colored egg-sacs to a leaf and then stand guard over them until the young hatch. This spider has the amazing ability to "spit" venom into the eyes of an aggressor. Similar species are found in southern Europe, and Africa.
DISTRIBUTION USA, from New England southward to Georgia and westward to the Rocky Mountains, and in Central America.

Sac Spiders *Family Clubionidae*
This is a family of more than 1,500 species of mainly brown nocturnal spiders. They spend the day in a silken tube built inside a rolled leaf or placed beneath a stone or log. They are active hunters and do not build prey-catching webs. Many species are superb mimics of ants.

CHEIRACANTHIUM ERRATICUM

COMMON NAME Grass-head sac spider
DESCRIPTION The legs are a translucent yellowish brown. The carapace is brown, and the velvety abdomen straw-colored, with a broad yellow central band bisected by a maroon median stripe. The females enclose themselves and their eggs in a nest, which is usually composed of a grass-head fastened together with silk to form a disk. *C. mildei*, found in southern Europe and North America (introduced) is very similar but is often found in buildings, where it has bitten humans, resulting in nasty skin-blisters.
DISTRIBUTION Europe and much of Asia.

CASTIANEIRA spp.

COMMON NAME Painted ground spider
DESCRIPTION More than 25 species of *Castianeira* are found in the USA. Many of these mimic ants, or are brightly colored and mimic velvet ants (mutillid wasps). Most

species raise and lower their abdomen and front legs as they scuttle around on the ground, rather in the manner of ants or wasps. The egg-sacs are in the form of a flattened disk with a metallic sheen and are usually fastened to the underside of a rock. The species illustrated is ³/₈ in/9 mm long and was in the Arizona desert. *C. occidens* from the deserts of southwestern USA is a striking species in which the abdomen is bright orange, and the glossy black carapace has a white median band.
DISTRIBUTION More than 25 species in the USA, many of them very widespread.

ZUNIGA MAGNA

COMMON NAME Ant spider
DESCRIPTION The top of the legs, thorax, and abdomen of this glossy black ½ in-/ 13 mm-long spider are partially clothed in a pelt of short golden hairs. Down the center of the abdomen there is a line of five white spots. The top of the head is very flattened, and the jaws project forwards. In common with most spiders that mimic ants, the front legs are held up and waved about as the spider walks around, thereby mimicking the constantly quivering antennae of an ant.
DISTRIBUTION Rainforests of eastern Brazil.

ANYPHAENA ACCENTUATA

COMMON NAME Buzzing spider
DESCRIPTION Both the female and the male (illustrated) are straw colored. In the male the sides of the abdomen and carapace are dark with four black spots in the middle of the rear half of the abdomen. The female is fatter and appears much paler, as her abdomen lacks the dark sides. This species is an active hunter on the foliage of trees, especially oaks. During courtship the male produces a buzzing sound by rapidly tapping the tip of his abdomen against a leaf. The females lay their eggs inside a nest consisting of several leaves fastened together with silk. Around 20 species of *Anyphaena* are found in the USA, most of which live under stones or logs, or among grasses and bushes.

DISTRIBUTION Occurs over Europe and Asia.

Stone Spiders *Family Gnaphosidae*
More than 2,000 species of mainly brown or black stone spiders are found worldwide.
The front spinners are rather long and tubular, and are widely separated. The eight eyes are
very small, befitting spiders that are mainly nocturnal hunters relying on touch and smell,
rather than sight, to locate prey. Most species spend the day under logs or stones, hence
the common name, emerging at night to hunt and mate.

SCOTOPHAEUS BLACKWALLI

COMMON NAME Mouse spider
DESCRIPTION With its furry brown
appearance and darting movements this
³/₈ in-/9 mm-long spider rather resembles a
small mouse. The fur has a somewhat greasy
sheen and the dark brown carapace is
narrowed at the front end. The spinners
protrude from the rear of the abdomen as
two small knobs. The mouse spider is
usually seen wandering slowly around on
the walls of houses, searching for prey.
The female in the illustration is guarding
her white egg-sac under a stone.
DISTRIBUTION Europe and most of Asia,
and in North America.

DRASSODES LAPIDOSUS

COMMON NAME Stone spider
DESCRIPTION The carapace of the ⁹⁄₁₆ in-
/14 mm-long female (illustrated) is a
grayish- to reddish-brown, with a black
marginal line, and is densely carpeted with
short hairs. The plump, egg-shaped
abdomen is similar but often has a dark
band down the middle, (this band is absent
in the spider illustrated). The jaws are shiny,
brownish black, and rather large. The male
is similar to the female, but ⅛ in/3 mm
shorter. The spider spends the day in a
silken sac or tube beneath a stone or loose
bark, emerging at night to hunt insects.
There are six species of *Drassodes* in the
USA. The common *D. neglectus* is a yellow
or light gray spider with indistinct chevrons

on the rear half of the abdomen.
DISTRIBUTION Europe, eastward to Japan,
and in North Africa.

ZELOTES APRICORUM

COMMON NAME Black zipper
DESCRIPTION This glossy black spider is one
of many similar-looking ¼ in-/6 mm- to
⅜ in-/9 mm-long species found in Europe
(more than 50 species in France alone),
and the USA (around 30 species, many of
them dark brown rather than black). The
carapace is noticeably narrowed towards
the front, so that the eyes are crowded into
rather short rows. These spiders live under
logs and stones, emerging at night to hunt.
DISTRIBUTION Europe and most of
temperate Asia.

Huntsman Spiders *Family Sparassidae*
This is a mainly tropical family of generally large, flattened spiders with long legs. They are often called giant crab spiders, because their legs are held out crablike at their sides. Unlike true crab spiders (Thomisidae) they have teeth on their jaws. Sometimes the term Heteropodidae is used for this family.

MICROMMATA VIRESCENS

COMMON NAME Green meadow spider
DESCRIPTION The female (illustrated) reaches a length of ⁹/₁₆ in/14 mm and often sits and waits on leaves to ambush prey. The legs and carapace are a deep bright green, but the abdomen is yellowish green, and there is a deeper green cardiac band near its base. The male only measures ³/₈ in/9 mm and is a very beautiful spider, quite different from the female. His carapace and legs are a rather somber dull green but his slim abdomen is gold on top with a broad, red central band and red sides.
DISTRIBUTION Europe, eastward through Asia to Japan.

HOLCONIA IMMANIS

COMMON NAME Giant huntsman
DESCRIPTION This large spider is common in houses but it also lives on trees in forests. The overall color is grayish brown, with the

legs banded dark and light, and a deep brown cardiac stripe extending from the front of the abdomen to about halfway down its length. It is a powerful hunter, and will tackle large insects, other spiders, and even lizards. Prey up to 4 in/10 cm long is taken by the 1⅜ in-/34 mm-long females. The male is smaller (1 in/25 mm) and lives with the female for a while during the mating season. Humans are occasionally bitten but the bite is not dangerous, although rather painful.
DISTRIBUTION Native to Australia but introduced to New Zealand.

PANDERCETES GRACILIS

COMMON NAME Lichen huntsman
DESCRIPTION The flattened body is mottled gray and brown, while the legs are adorned with flattened tufts of hairs. The coloration helps the spider to blend into the lichen-speckled tree bark on which it spends the day, while the hairiness of the body and legs helps to eliminate shadows. The female (illustrated) reaches a length of ¾ in/19 mm, and guards her flattened disk-shaped egg-sac. She makes an attempt at camouflaging this by incorporating a few flakes of lichen into the tough silk on its surface.
DISTRIBUTION Australia (tropical Queensland) and New Guinea.

Wall Crab Spiders *Family Selenopidae*

This is a small family containing some 400 species of very flattened, gray or brown spiders which live under stones or on rocks. The arrangement of the eyes, six in a single row, is characteristic of the family.

SELENOPS RADIATUS

COMMON NAME Wall crab spider

DESCRIPTION This is one of only four species of the genus found in the USA. The spider illustrated was found in the Arizona desert. The body and legs are heavily mottled in dark and light brown, making the spider very difficult to see when it flattens itself against some desert rock-face. While waiting for a meal to turn up the spider sits head-downward but will sprint toward a crevice with remarkable speed if disturbed. Several species are common in houses, living behind picture frames, fridges, furniture or toilets.

DISTRIBUTION Widespread in warm regions, including southern USA.

Two-Tailed Spiders *Family Hersiliidae*
This family contains some 75 mainly tropical spiders which live on tree trunks and rock-faces. They are easily recognized by the rather triangular abdomen, and the pair of long spinners which protrude from the rear end, like two tails.

HERSILIA BICORNUTUS

COMMON NAME Two-tailed spider
DESCRIPTION As with all two-tailed spiders, the brownish or grayish mottled coloration makes the spider very difficult to spot when it is sitting head-downward on a tree trunk. The lens-shaped egg-sacs are covered in white silk, which the female camouflages with bits of bark or lichen prized off with her chelicerae. When an insect passes by, the spider fastens it to the trunk by running in circles around it, trailing silk from the long spinners. These spiders can often be found in the less dense parts of forests, such as along highways or around the edges of clearings.
DISTRIBUTION Southern Europe and Africa.

Crab Spiders *Families Philodromidae and Thomisidae*
The spiders in these two families are crablike in both shape and movement, walking forward, sideways or backward with equal facility. The males are much smaller than the females but have much longer legs in proportion to their bodies. Crab spiders do not build prey-catching webs but instead sit in ambush on flowers, leaves, bark, rocks or soil. More than 2,000 species have been described, mainly from the tropics.

PHILODROMUS DISPAR

COMMON NAME House crab spider
DESCRIPTION This philodromid can often be seen wandering around inside houses. It is also common on trees in woodland and on shrubs in gardens. The male (illustrated) is a slim-bodied spider with an iridescent black carapace and abdomen, and white sides. The female looks like a completely different species. She has a plump brown abdomen, and a brown carapace with dark bands on either side, bordered by white margins. She can often be found with her legs spread out protectively across her white silken egg-sac tucked into the corner of a wall.

DISTRIBUTION Europe, through most of Asia, and in North America.

TIBELLUS OBLONGUS

COMMON NAME Grass spider

DESCRIPTION This long, thin philodromid often sits head-downward on grasses, where it is very hard to spot. The body and legs are straw colored, and there is a light brown band down the middle of the body from front to rear. A short way in from the tip of the abdomen there is a pair of black spots.

The females measure ³/₈ in/9 mm, the males slightly less. The females can often be found sitting astride their silk-covered egg-sacs affixed to a plant stem.

DISTRIBUTION Europe, eastward to Japan, and in most of North America.

XYSTICUS GULOSUS

COMMON NAME Plain crab spider
DESCRIPTION This is one of more than
70 species found in the USA, with another
17 in Europe. Nearly all of these are
rather similar in size and coloration to the
$^5/_{16}$ in-/8 mm-long female in the illustration.
In many species the abdomen is more
heavily marked with dark chevrons, while
one or two are spotted. In some species the
tiny, dark brown males spin a bridal veil
over the female's head and legs before
mating. The females stand guard over their
white egg-sacs attached to plants. Although
often found on flowers, most species of this
genus spend much of their time waiting in
ambush on leaves, bark, stones or sand.
DISTRIBUTION Most of the USA, excluding
the deserts of the southwest.

THOMISUS ONUSTUS

COMMON NAME Heather spider
DESCRIPTION The females of this handsome spider can be pink, pale yellow or white. The pink form is almost always found on pink flowers, especially heathers, but the white and yellow forms do not always choose a matching background. The female (illustrated) has a very plump and rather triangular abdomen with angular tubercles on the top corners. On the face there are two horn-like projections on either side of the eyes. The females reach a length of $^5/_{16}$ in/8 mm but the brownish-orange males are tiny.
DISTRIBUTION Europe, eastward to Japan, and in North Africa.

THANATUS FORMICINUS

COMMON NAME Diamond spider
DESCRIPTION The ground color of this long-legged philodromid is light brown. At the front of the abdomen there is a rather elongate, dark brown, diamond-shaped mark, edged with white. Down the center of the carapace there is a broad, pale brown band. The male measures $^1/_4$ in/6 mm in length, the female $^7/_{16}$ in/11 mm. The diamond spider hunts for prey on tree trunks, cliff-faces, and among grasses, and shrubs.
DISTRIBUTION Europe, eastward through Asia, and throughout most of North America.

MISUMENA VATIA

COMMON NAME Common flower spider
DESCRIPTION In the USA this spider is often known as the goldenrod spider from its habit of sitting on the yellow flowers of that name. The same individual is just as likely to be found on white flowers, because this spider can change from yellow to white and back again, according to the color of the flowers it chooses as a background. The sides of the abdomen are often striped with red. The small, dark brown male cannot change color. The venom is very potent and insect victims rapidly succumb to its effects.
DISTRIBUTION Europe, eastward to Japan, and in North Africa and North America.

PHRYNARACHNE RUGOSA

COMMON NAME Warty bird-dropping spider
DESCRIPTION The wrinkles and warts which cover this strange spider, allied to its shiny appearance and blotchy coloration, all conspire to make it a perfect mimic of a wet, recently fallen bird-dropping. The ³/₈ in-/ 9 mm-long female (illustrated) sits motionless for day after day in full view on the same leaf. She does not merely wait passively for prey to chance by, but actually emits a special manure-like scent that attracts certain types of flies to come within reach of her grasping front legs.
DISTRIBUTION Tropical Africa and Madagascar.

SYNEMA GLOBOSUM

COMMON NAME Gold leaf crab spider
DESCRIPTION The small, shiny, and rather rounded abdomen of the female can be black and red (as illustrated), black and gold, or black and yellow. The black pattern on the abdomen resembles a human face. The carapace is black, which contrasts with the pale brown eyes. The females reach a length of ¼ in/6 mm and are usually found on flowers. The small black males have a white bar on the abdomen. There are three species of *Synema* in the USA. In *S. parvula* the legs and carapace are yellowish orange and the abdomen yellow with a black tip. It is common in the eastern states.
DISTRIBUTION From southern Europe through temperate Asia to Japan, and in North Africa.

CAMARICUS FORMOSUS

COMMON NAME Hallowe'en crab spider
DESCRIPTION The rather domed carapace is deep red, contrasting with the milky white and rather ghostly eyes. The glossy abdomen is longer than broad, with a black Hallowe'en-type mask set against a white background, and a red rear end. The legs are a translucent grayish white, with a few black markings. The female (illustrated) reaches a length of ³/₈ in/9 mm. This is one of a number of small, glossy crab spiders that prey solely on ants.
DISTRIBUTION India to Java, and in Vietnam.

STEPHANOPIS ALTIFRONS

COMMON NAME Knobbly crab spider
DESCRIPTION The rough exterior of this rather square-bodied crab spider perfectly matches the texture of the bark on which it normally lives. It often wedges itself into a crack in the bark, making itself even less conspicuous. The general coloration is grayish brown. At night the spider comes alive and wanders over the bark in search of prey. The female, which reaches a length of ⁷/₁₆ in/11 mm, hides her egg-sac inside a crevice in the bark. Like the lichen huntsman and two-tailed spiders, she camouflages it with flakes of bark.
DISTRIBUTION Eastern Australia.

Jumping Spiders *Family Salticidae*

This is a huge family of more than 5,000 species, most of which are tropical. Many of the males sparkle with iridescent, jewel-like colors, and often look quite different from the much drabber females. Most species are rather small, the largest being only ⁷/₁₀ in/35 mm long. They have eight eyes. Two large ones face forward and can be focused very accurately from as far as 20 in/50 cm away. The other eyes are smaller, and help to detect movement and fix the prey's position. These spiders can leap up to 30 times their own length.

SALTICUS SCENICUS

COMMON NAME Zebra spider

DESCRIPTION This ¹/₄ in-/6 mm-long, black and white spider is usually seen stalking across the walls and windowsills of houses. It is most common around houses and backyards but is also at home on rocks and trees away from habitations. The female (illustrated) stands guard over her egg-sac, which is normally placed under a rock, plank, flowerpot or other object. The carapace is black with two white spots, while the black abdomen has a white frontal margin and two pairs of white, backward-pointing chevrons. The male's jaws are large and project well forward of the face.

DISTRIBUTION Europe, through most of Asia, and over most of North America.

PLEXIPPUS PAYKULLI

COMMON NAME Pantropical jumper
DESCRIPTION The very broad, square face of this attractive little ⁷/₁₆ in-/11 mm-long spider is marked with horizontal black and white stripes. The carapace is mainly black, with a pale central stripe, and the abdomen is black with one or two central stripes, sometimes partly broken up into dots and with white edges. Jumping spiders will often tackle prey far larger and fiercer than themselves, as depicted in the illustration, which shows a spider feeding on a katydid.
DISTRIBUTION In most of the warmer parts of the world, including southern USA (Georgia and Florida westward to Texas).

ERIS AURANTIUS

COMMON NAME Variable jumper
DESCRIPTION This species is very variable. The body is heavily covered in iridescent

scales, so that the colors appear different depending on the angle. The overall color is an iridescent, blackish bronze. Each side of the carapace is marked with a whitish-orange band. The entire margin of the abdomen bears a broad, orange band, although this can sometimes be very pale, as in the individual illustrated. On top of the abdomen there are two or three pairs of white spots, and usually two pairs of orange spots, which are less conspicuous in the male. Females reach a length of ⁷/₁₆ in/11 mm; males are marginally smaller.
DISTRIBUTION Central America and southern USA, from Florida to Arizona and northward to Delaware and Illinois.

LYSSOMANES VIRIDIS

COMMON NAME Leaf jumper
DESCRIPTION The ⁷/₁₆ in-/11 mm-long female (illustrated) is broadly similar to the male. The legs and carapace are vivid green, with the eyes set within a rectangular patch containing stripes of red, brown, and white scales (sometimes just light brown). The slender, tapering abdomen is of a more brownish shade of green. The males have huge curved chelicerae which jut out well forwards of the face and are overtopped by the even longer stalks of the palps.
DISTRIBUTION Central America and the Caribbean, and southeastern USA (North Carolina southward to Florida and westward to Texas).

PHIDIPPUS JOHNSONI

COMMON NAME Johnson's jumper
DESCRIPTION The carapace is black in both sexes of this attractive spider. In the males the abdomen is a strikingly bright red, with a black hind margin. In the ⅜ in-/9 mm-long female (illustrated) the red is more subdued, and only occurs as a band down the center of the black abdomen, or along its margins. The face has a small mustache of white hairs. The legs are black, and in the males they are adorned with substantial tufts of hair.
DISTRIBUTION Over much of southern USA, from North Dakota southward to Texas and then across to the Pacific coast states.

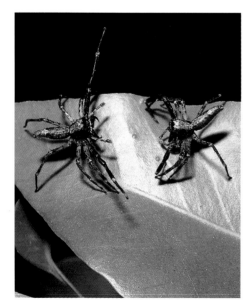

HELPIS MINITABUNDA

COMMON NAME Bronze Aussie jumper
DESCRIPTION In both sexes the abdomen is long and slender, and the carapace broad and high. The males are an iridescent, bronze-brown, with very long black legs with white joints. These long legs are used in competitions between males, as depicted in the illustration. The female is paler and has shorter front legs. She is often found in a nest, consisting of a silk sheet forming a roof over a leaf.
DISTRIBUTION Eastern Australia and New Zealand.

TELAMONIA DIMIDIATA

COMMON NAME Two-striped gaudy jumper
DESCRIPTION The male (illustrated) reaches a length of nearly ½ in/13 mm. His black front legs are adorned with dense fringes of black hairs. There is a red band across the eyes, and a white spot on top of the black carapace, which has broad white sides. The slender, tapering abdomen bears broad, lengthwise bands of blackish-red and white. The slightly longer female looks completely different, with a pale whitish-orange ground color and two narrow orange stripes along the top of the abdomen. The carapace bears a number of orange blotches.
DISTRIBUTION India to Indonesia.

Acari 7
Achaearanea tepidariorum 27
Aculepeira ceropegia 41
African golden orb-weaver 37
Agelena labyrinthica 33
Agelenopsis aperta 33
Amaurobius 25
Amblypygi 8
American green lynx spider 59
American nursery-web spider 58
Aname 13
ant spider 61
Anyphaena accentuata 61
Aphonopelma chalcodes 15
Arachnida 7
Araneae 10
Araneomorphae 10, 12, 16–79
Araneus 38–40
Araniella cucurbitina 42
Arctosa 56
Argiope 28, 43–6
Argyrodes 28, 37
Ariadna bicolor 21
Arizona black hole spider 26
arrow-shaped thorn spider 53
arthropods 6–7
Atrax robustus 13
Atypus affinis 12
Australian trap-door spider 13

baboon spider, common 15
banded argiope 46
bankside orb-weaver 42
basilica spider 53
bird-eating spiders 14–15
black lace-weaver 25
black-striped orchard spider 36
black widow spider 29
black zipper 63
blunt-spined kite spider 50
bordered orb-weaver 41
Brachypelma smithi 14
bridal-veil lynx spider 59
bronze Aussie jumper 78
brown spiders 17
Bruennichi's argiope 45
buzzing spider 61

Camaricus formosus 74
Carolina wolf-spider 55
Castianeira 60
cellar spiders 18, 29
Cheiracanthium erraticum 60
chelicerae 8–9, 12
Christmas spider 52
Chrysso 30
cobweb spider 34
cobweb-weavers 27–30
comb-footed spiders 27–30
conical orb-weaver 48
crablike spiny orb-weaver 51
crab spiders 9, 68–74
cribellates 9, 23–4
cross spider 38, 39
Cyclosa 48
Cyrtophora hirta 47
daddy long-legs spiders 18

dainty platform spider 32
Deinopsis 22
desert bush spiders 19
desert grass spider 33
dewdrop spiders 28, 37
diamond spider 71
diet 8–9
Diguetia canities 19
Dolomedes 58
Drapetisca socialis 32
Drassodes 63
dwarf spiders 31–2
Dysdera crocata 20

earth-chaser 54
egg-sac 9
Enoplognatha ovata 28
Eris aurantius 76
European fishing spider 58

feather-footed spiders 23
fishing spiders 58
flower spider, common 72
four-spot orb-weaver 39
funnel-weavers 33–4
funnel-web tarantulas 13
furrow spider 42

garden spider 38, 39
Gasteracantha 49–52
giant huntsman 65
glasshouse spiders 27
golden orb-weaver 37
golden-silk spider 37
gold leaf crab spider 73
goliath tarantula 14
grass-head sac spider 60
grass spider 69
grass spiders 33–4
great golden argiope 43
green meadow spider 64
green orb-weaver 42

habitats 6
hairy tent-spider 47
Hallowe'en crab spider 74
Harpactira gigas 15
harvestmen 6, 7
Hasselt's spiny spider 50
heather spider 71
Helpis minitabunda 78
Herennia ornatissima 46
Hersilia bicornutus 67
Holconia immanis 65
hole spiders 26
horned orb-weaver 51
house crab spider 68
huntsman spiders 64–5
invisible spider 32

jaws *see* chelicerae
Johnson's jumper 78
jumping spiders 75–9

knobbly crab spider 74
Kukulcania arizonica 26

lace-weavers 24–5
Latrodectus 29
leaf jumper 77
leaf lace-weaver 24

leopard spider 21
Leucauge 35–6, 53
lichen huntsman 65
lobed argiope 45
long-horned orb-weaver 49
Loxosceles 17
lumpy thorn spider 52
Lycosa carolinensis 55
lynx spiders 59
Lyssomanes viridis 77

mabel orchard spider 36
marbled orb-weaver 40
Mecynogea lemniscata 53
Metellina segmentata 36
Mexican red-knee tarantula 14
Miagrammopes 23
Micrathena 52–3
Microlinyphia pusilla 32
Micrommata virescens 64
missing sector orb-weaver 48
Misumena vatia 72
mites 7
molts 7
money spiders 31–2
mountain orb-weaver 41
mouse spider 62
Mygalomorphae 10, 12–15

Neoscona adianta 41
Nephila 28, 35, 37
Neriene peltata 31, 32
net-casting spiders 22
Nigma puella 24
Nuctenea cornuta 42
nursery-web spiders 57–8

Opiliones 7
opisthosoma *see* abdomen
orb-weavers 38–53
orchard spiders 36, 53
ornate orb-weaver 46
Oxyopes schenkeli 59

painted ground spider 60
palps *see* pedipalps
Pandercetes gracilis 65
pantropical jumper 76
Pardosa 55
Peucetia viridans 59
Phidippus johnsoni 78
Philodromus dispar 68
Pholcus phalangioides 18
Phrynarachne rugosa 72
Physocyclus 18
pigmy mesh-spinners 24
Pirata piraticus 56
pirate spider, common 56
Pisaura 57
plain crab spider 70
platform-web spider 31
Plexippus paykulli 76
prey 8–9
prosoma *see* cephalothorax
purse-web spiders 12

recluse spiders 17
red and silver dewdrop spider 28
red and white cobweb spider 28

sac spiders 60–1
Salticus scenicus 75
sand-runner 56
Scorpiones 7
scorpions 7
Scotophaeus blackwalli 62
Scytodes thoracica 16
Segestria senoculata 21
Selenops radiatus 66
shamrock spider 39
silver argiope 44
six-eyed spiders 20
six-spotted fishing spider 58
snake-back spider 21
Solifugae 7
Sphodros rufipes 12
spinners 9
spitting spiders 16
spotted wolf-spider 55
Steatoda grossa 29
Stephanopis altifrons 74
stick spider 23
stone spiders 62–3
Sydney funnel-web spider 13
Synema 73

tailless whipscorpions 8
tarantulas 14–15
Tegenaria 18, 34
Telamonia dimidiata 79
Tetragnatha extensa 35
Texas orb-weaver 39
Thanatus formicinus 71
Theraphosa blondi 14
Theridion melanurum 27
Thomisidae 9
Thomisus onustus 71
thorn spiders 52–3
Tibellus oblongus 69
triangular spine-leg spider 30
Trochosa terricola 54
tube-web spiders 21
two-striped gaudy jumper 79
two-tailed spiders 67

Uropygi 7–8

variable jumper 76
velvet mite 7
venusta orchard spider 36
violin spider 17

wall crabs spiders 66
warty bird-dropping spider 72
wasp spider 45
webs 9
wedding-present spider 57
Western desert tarantula 15
whipscorpions 7–8
window lace-weaver 25
windscorpions 7
wolf spiders 54–6
woodlouse spider 20

Xysticus gulosus 70

yard spider 34

zebra spider 75
Zelotes apricorum 63
Zuniga magna 61
Zygiella x-notata 48